This book is dedicated to all
math teachers
and parents out there who look
for rich literature with which
to educate our children!

Once upon a time, in the **Medieval** days, there was a small kingdom in the Provice of Operations.

("Medieval" means hundreds of years ago — in case you didn't know).

This tiny kingdom was named Math-Ville and was run by respected King and Queen, Peter and Penelope Parentheses.

During Midieval times, these kinds of kingdoms consisted of only a hundred or so people, so it was traditional for dinner mealtimes to be held at-court.

These meals were shared events in which the king and queen presided over the food with their faithful people along with their exalted son and daughter, the prince and princess.

Now the king and queen were most proudly served by The Grand Duke, Edward Exponent.

He was in charge of making sure the royal chefs got all the meals right and that the people (including Prince Manny Multiplication and Princess Deborah Division) were served in the correct order — and according to their station.

In those days, the Grand Duke, Edward Exponent, made sure the people were all served before *he*, himself received his meal for the evening.

He convinced himself that being last was his due, with respect to his station of looking after the best interests of the kingdom.

However, there were some *serious* problems…

First of all, Prince Manny Multiplication and Princess Deborah Division were in a *constant* **quarrel** over who was to be served *first!*

("Quarrel" means argument — in case you didn't know).

The Prince believed that *he* should *always* be served first, after the king and queen of course, as *he* was the first born and the heir to the throne of Math-Ville.

The princess disagreed, as *she* was often the *first* one ready and downstairs to greet guests as they entered the castle for their meal.

She thought *she* should be served first if she was downstairs first, and *always* before the common people, the attractive-addition men and the sophisticated-subtraction women.

The Grand Duke, Edward Exponent, grew tired of their bickering, but he was also tired of the bickering of the common people.

The attractive-addition men believed that *they* should be served before the sophisticated-subtraction women because *they* were the ones who provided for their families.

After all, they were served morning and noon meals *by* the women who cared for them and took care of their baby pluses and minuses.

The sophisticated-subtraction women argued that *they* should be served first! They declared that if the attractive-addition men *had* no sophisticated-subtraction women to care for them, *they* would be left to starve!

One long summer day, when the arguing had finally become so intense that a food fight was breaking out by the salad cart, the Grand Duke finally decided he had *enough!*

As King and Queen, Peter and Penelope Parentheses, looked on, the Grand Duke calmly but **staunchly** sat down with his plate and began eating his meal before *any* of the common people or *even* the prince and princess had been served!

("Staunchly" means firmly and deliberately – in case you didn't know).

Prince and Princess, Manny Multiplication and Deborah Division, stared at one another, **aghast** at the duke's bold move!

("Aghast" means shocked and horrified — in case you didn't know).

They were *so* overcome, that they stopped their manhandling of the meatloaf to look to their mother and father, King and Queen, Peter and Penelope Parentheses, with pleading eyes for them to stop this disgrace!

As the common people saw their royal prince and princess stop fighting, they dropped the food they were prepared to fire in shame.

The king and queen decided it was *finally* time to come off their matching thrones and address the situation. A hush fell over the crowd as they approached.

"As ruler of this people," King Peter Parentheses declared, "I am *ashamed* to call *any* of you my own - *including* my own son and daughter!

"You spend *so* much time fighting over who will be first to eat the food my queen and I provide you, that it no longer becomes a pleasure to dine with *any* of you!"

Prince Manny and Princess Deborah stared in disbelief at their father and mother. "But she...!" Manny Multiplication began before he was cut off with a scathing look from his mother.

"But he…!" Princess Deborah Division cried out before her father gave her a glare that could frighten a dragon.

The Grand Duke, Edward Exponent, calmly stood, brushing of the remains of his meal, and looked at the people of Math-Ville. Then he looked at the king and queen for their next move.

It was possible that he could be beheaded for defying the orders of the king and queen, but he was simply too *tired* of all the mayhem, he almost no longer cared.

However, instead of **reprimanding** him, the king and queen calmly stood united at the head of the table.

("Reprimanding" means scolding — in case you didn't know).

In a loud voice that could be heard clear across the Province of Operations, the king boldly declared:

"It shall be known, from here on, that Grand Duke, Edward Exponent, for his grace and dignity in all of this chaos, shall be the *second* to eat behind the King and Queen Parentheses!"

There was a hush and a gasp among the crowd, including the royal children who still held food, ready to launch at each other.

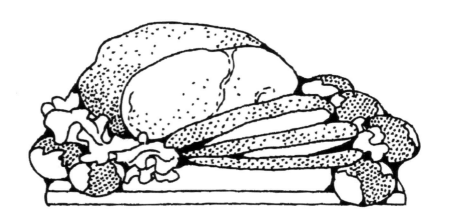

The King continued. "Furthermore, although Prince Manny Multiplication and Princess Deborah Division are *equal* in our eyes, whichever of them arrives down to the meal *first* will be the one served *first!* And that goes for **all** of the people as well!"

He cleared his throat and looked pointedly at the crowd with their jaws dropped.

"Concerning the attractive-addition men and the sophisticated-subtraction women, whichever of you arrives in line to be served *first* will be *before* the other!

"But you will *never* be before the queen and I, King and Queen Parentheses, before Grand Duke, Edward Exponent, or *either* of my two royal children, Prince Manny Multiplication or Princess Deborah Division!

"Am I understood?"

The people, including the royal children, looked ashamed.

The Grand Duke, however, looked thrilled and relieved. He still had his head, after all, and he was now *second* after the king and queen and always *before* the prince and princess!

The queen then calmly came and laid her hand in the King's and declared: "So be it!"

After the fateful evening that decided the order of the people of Math-Ville in the Province of Operations, the bickering stopped, and Prince Manny Multiplication and Princess Deborah Division began to try to beat each other *fairly* down to dinner.

They knew whoever came down *first* would determine who ate first and who was on the left and right side of the long banquet table. Whichever came down *first* sat on the left, and the later one was on the right.

Similarly, the people of the land began respectfully gathering in a line, from left to right, so they could still take part in the meal according to who came first (after the king and queen, the grand duke, and the prince and princess, of course!).

Finally, order was restored to the kingdom of Math-Ville, peace returned, and meal times within the kingdom became pleasant events.

The king and queen who reigned with fairness became proud of their children and their people once more.

And the Grand Duke, Edward Exponent, continued to eat *most* of his meals at the table with the royal family. From then on, they *all* lived happily ever after.

*** THE END ***

***A final note for all my math students who might argue that sometimes Edward Exponent comes first:

$$6 \times (5 + 4^2 - 2^3)$$

That **only** happens when he sneaks his food **inside** the King and Queen Parentheses' royal court and has to sneak out before being caught (lest he wish to lose his head)!***

Made in the USA
Middletown, DE
24 February 2025

71811872R00020